Concerning
BORAX BILL
AND THE FAMOUS
20 MULE BORAX
TEAM from
Death Valley
California

PACIFIC COAST BORAX CO., U. S. A.

BORAX BILL.

The 20 MULE BORAX TEAM *and* ITS FAMOUS DRIVER

BEAUTIFUL WOMEN.

THE beautiful women of old Egypt used Borax to keep their flowing garments soft and snowy In Pompeii the ladies of the nobility used Borax in the public baths and they were famed in the ancient world of culture for their sweetness and charm. Borax was known and used by the elect in the Seventh Century but it remained for the Wonderful West of America to produce this valuable mineral in quantities large enough to enable all to have the great advantage of its use. The first large deposits in modern days were found in 1872 in Nevada and Death Valley, California. The largest producer of borax and borax products in the world, the Pacific Coast Borax Co., are refining the output of these deposits.

To bring greater knowledge of borax to all households the famous 20 Mule Borax Team is now touring America.

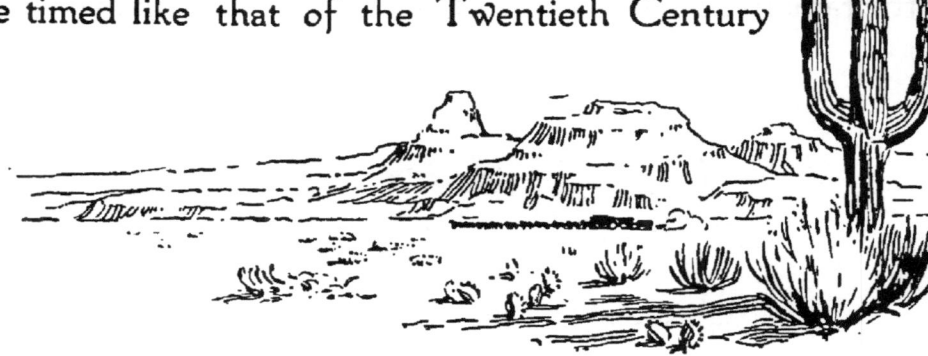

20 MULE TEAM.

Transportation of the borate mineral from the mines in Death Valley to the railroad at Mojave, California, a distance of 162 miles, made necessary the now justly celebrated 20 Mule Borax Team.

The journey, hot, tedious, long, dusty and tortuous, wound over the parched and drifting sands of the desert and through the dry and rocky ravines of the Funeral Mountains. Twenty days were consumed in making the round trip and there was not a single house or habitation on the entire route. One stretch of sixty miles was without water, consequently provisions and water for the round trip were loaded on the wagons at the railroad and cached for the return trip. Water, that precious fluid in the desert, was carried on the round trip in a steel tank containing 1,200 gallons. The teams ran on scheduled time and their arrival could be timed like that of the Twentieth Century

Limited.   They were just as sure but not quite so fast.   A 20 Mule Team hauled 30 tons of borax. This capacity is considered remarkable in view of the fact that the modern railway car on steel rails has only a twenty ton capacity.   The mule tractors hauled their tremendous loads up and down the rocky canyons and steep grades of the Funeral Mountains and over the burning sands of Death Valley and there is no record of even one break down.   If you were on the lookout for a Borax train to come in, you would first see, far off in the desert, a small cloud of dust, white as steam, slowly growing larger until finally the outlines of huge wagons could be discerned; later the tinkling of the chime bells on each mule.   The rancus roar of Borax Bill shouting his commands—the distant rumble of the mammoth wagons growing louder and louder, the thump of eighty hoofs—the creak of chains, the heavy breathing of this small army of mules and—presto—there is the famous animated trade mark, the 20 Mule Borax Team right before your eyes.

The driver rides on the "nigh-wheel horse." A large strap is attached to the end of the brake bar, and with a snap hook fastened to a ring in the back of the driver's saddle, the brake bar can be thrown in a ratchet on the side of the wagon, thus keeping the brake on until it is thrown out. The driver swings himself into the saddle, reaches back and pulls taut the brake strap, at the same time signalling with the "jerk-line" to the leaders, giving a peculiar call to the team. Immediately the animals "wake-up," and "get into their collars," and tighten up the long chain that reaches from the leader to the wagon; the driver gives another peculiar shout to the team, at the same time giving the brake strap a tremendous pull, releasing the break bar from the ratchet; the animals all pull in unison, the brake is off, and the great desert caravan is under way.

The "20 Mule Team" has become the Trade Mark of the Pacific Coast Borax Company's products, and is recognized the world over as a guarantee of purity. This team has been a unique feature

in the history of the West. As a method of transportation it has been out-grown by the continually increasing demand for Borax in the home, but it will always be associated with Death Valley, the home of Borax.

The present tour of the "20 Mule Borax Team" might be termed as a revival of a classic. The Pacific Coast Borax Company has always been noted for its enterprise and originality in advertising campaigns, and no more striking or effective means of bringing to the attention of the public the merits of the famous 20 Mule Team Borax products could be devised, than the restoration of the "20 Mule Borax Team" in its original form making a tour of the country. This may be the last opportunity for the public to witness this interesting and educational and thrilling historical spectacle.

Mules, like people, vary in intelligence. For service in a 20 Mule Borax Team only animals of extra-ordinary intelligence can

be used. Each animal must know his name and promptly obey commands addressed to them. He must be strong, and willing to do his share of the work. The most important and valuable animal in the team is the " nigh leader." The swinging of the team in rounding curves and the like depends greatly upon his doing his work intelligently. Each span of mules is attached to a

THE FAMOUS 20 MULE BORAX TEAM

set of single-trees and a double-tree, hooked into the chain which extends from the leader to the wagon. In going around a sharp curve, naturally this chain would be on a tangent from the leaders to the wagon, there-fore, in order to keep the chain in this periphery of the curve, as well as the wagon in the road, it is necessary to have some of the spans of mules between the leaders and the wagon leap over the chain, and pull almost at right angles to the direction of the team, compelling them to step along "sideways." This they will do upon the driver shouting the command to them by name. At the end of the day's journey the mules are unhooked from the single trees, and the chain, with the double and single trees attached, is left

ON
TOUR
FROM
COAST
TO
COAST

at full length in front of the wagons. In the morning the driver and "swamper" put on the harnesses, hanging the bridles on the hames, and each mule after drinking takes his own place at the chain, although there is no special mark to indicate it, and they are not camped two consecutive nights at the same place.

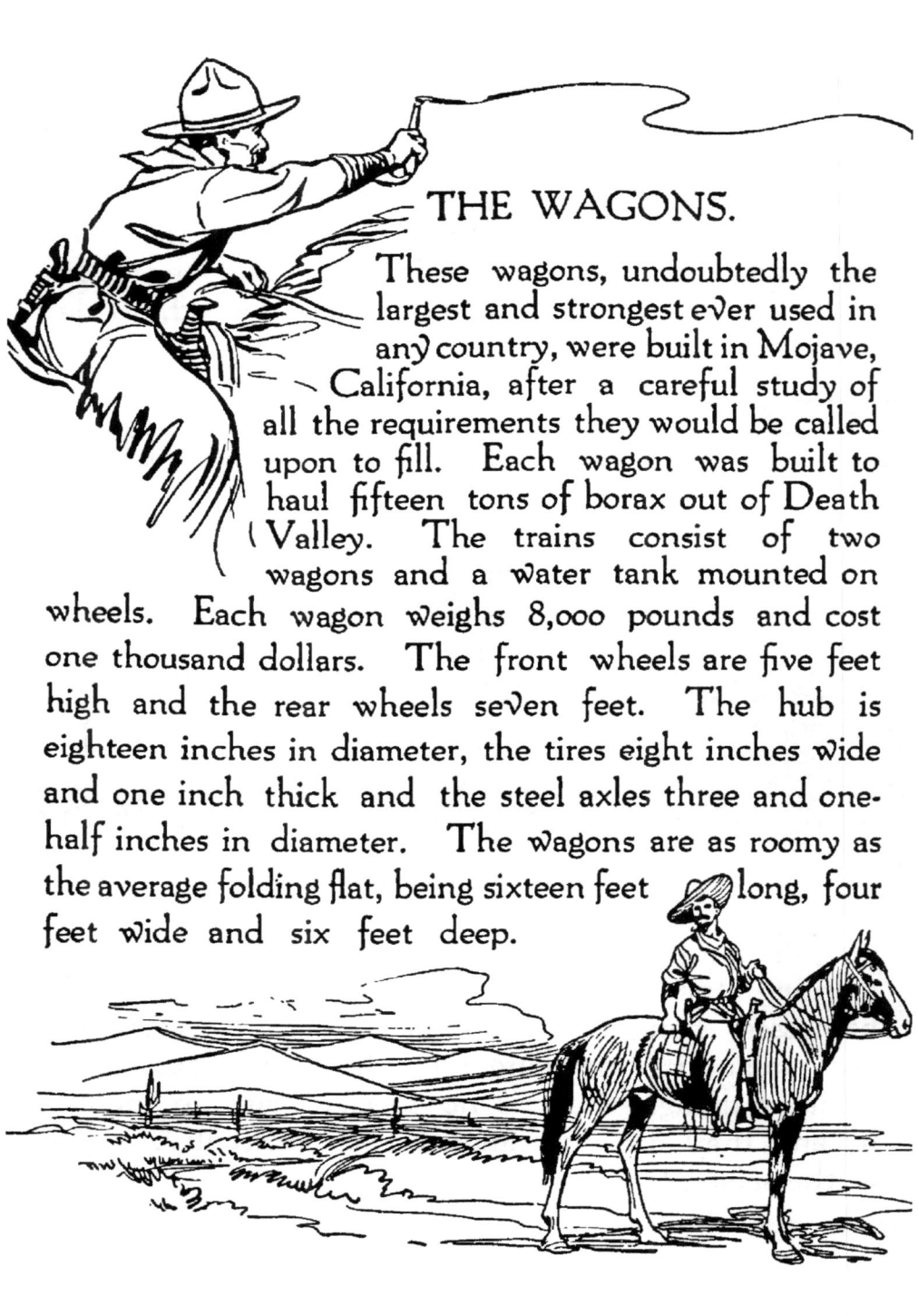

THE WAGONS.

These wagons, undoubtedly the largest and strongest ever used in any country, were built in Mojave, California, after a careful study of all the requirements they would be called upon to fill. Each wagon was built to haul fifteen tons of borax out of Death Valley. The trains consist of two wagons and a water tank mounted on wheels. Each wagon weighs 8,000 pounds and cost one thousand dollars. The front wheels are five feet high and the rear wheels seven feet. The hub is eighteen inches in diameter, the tires eight inches wide and one inch thick and the steel axles three and one-half inches in diameter. The wagons are as roomy as the average folding flat, being sixteen feet long, four feet wide and six feet deep.

THE JERK-LINE.

The jerk-line is one of the most important pieces of the 20 Mule Borax Team equipment. It reaches from the driver to the leaders, and is one hundred and twenty feet long, made of the best leather, soft, pliable and strong. It is the drivers telegraph line, by means of which he communicates his wishes to the leaders of the team. It is by this single line the entire team of 20 animals is managed. From the driver it is carried to the nigh leader, through a ring on the rump of each mule, and through the hame rings, or other rings fastened on the housing.

The jockey stick, which connects the leaders, is a light iron rod, with a snap hook on either end; one end is fastened on the chin strap of the nigh mule, the other to the hame ring on the off side of the off mule. The jockey stick guides the off mule. When the driver wishes the leaders to go to the right he gives a strong, steady pull; when he wishes the team to go to the left he jerks the line, the latter operation giving this line its name.

"BORAX BILL"

The man who is acknowledged to be the best handler of the jerk-line and driver of 20 mule teams, is William Frank Wilson, better known as "Borax Bill." From his youth he has been driving these mammoth teams, and has a copyright on a large number of peculiar expressions that he has found necessary in awakening the required amount of energy in balky mules. Bill understands every word of mule language; talks to them in their own tongue, and uses many expressions while handling these animals that would not sound well in polite society.

Mules are better than horses for freighting through a rough, mountainous country and as they are very careful about having a sure-footing, they seldom make a mis-step.

Mules have some very marked peculiarities, and at times an entire team will seem to be in a conspiracy to do everything contrary. Borax Bill soon discovers that fact; the team is brought to a stop, Bill deliberately dismounts, takes his black snake, which has in all probability been coiled in a necklace around his neck, and makes a few emphatic, eloquent

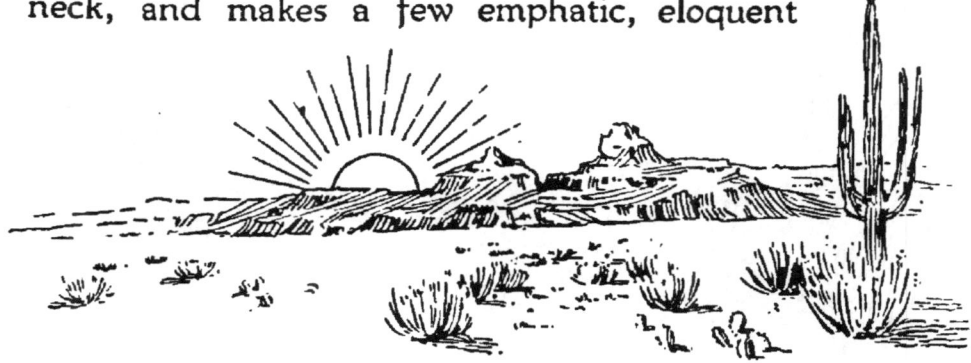

statements to Mule No. 1, accompanying his remarks with a round lot of blows from the whip, by way of being convincing. This he will repeat with each mule on both sides of the team. All the animals having been impartially attended to, Borax Bill again mounts and with a few pulls on the jerk-line notifies the leader that he is ready to move. All the animals are at once in readiness for the starting signal, which is promptly given.

They settle forward in their collars, the chain is pulled taut, off goes the brake, and with another command from Bill they are pulling in unison and with a will, and the mighty load of 85,000 pounds is moving, the crankiness of mules has disappeared, and the conspiracy for the time is at an end.

William Frank Wilson was born in 1858, and drove his first 20 Mule Team when he was twenty years old, after two or three years experience as a swamper, hauling merchandise and mining supplies from Carson City, Nevada to Bodie, California. In 1883 he went to Daggett, California, and drove a 20 Mule Borax Team ---hauling borax from Death Valley to the railroad at Daggett.

In freighting with a 20 animal team every driver has an assistant, called a "swamper."

The swamper's duties are numerous. He has to cook the food where they make their camp, wash the dishes and look out for fuel for the fire to cook the meals (this fuel usually consisting of sage brush or grease wood); in going up grade he has to get out and walk alongside of the team; on the down grade he operates the brake of the rear wagon; in camp he assists in unhooking and unharnessing the mules and in feeding them.

The building of railroads to all portions of the great West has made these mighty teams relics of the past. They have, however, performed an interesting and useful part in the service of man and the development of our country. Without them Death Valley would be but a reminder of a grewsome historical event.

DEATH VALLEY, CALIFORNIA.

This remarkable place came by its name in 1850. A party of immigrants, bound for California, attempted to use the desert as a shortcut to the gold field. They safely arrived at the edge of Death Valley, where they camped over night. In the morning they attempted to cross this narrow

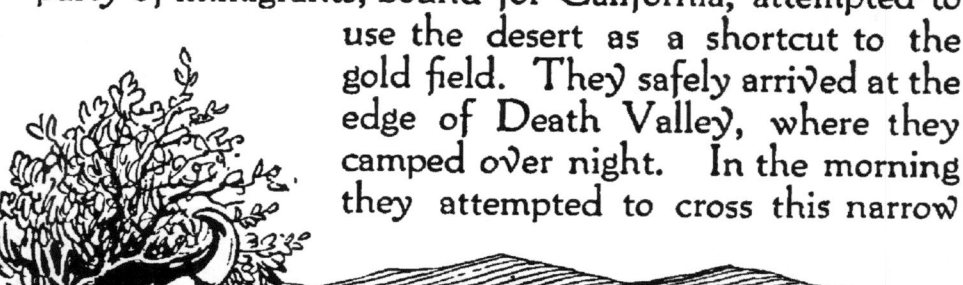

rent in the earth, whose floor is over 200 feet below the level of the sea, and where the dry air registers from 120 to 140 degrees. With the first streak of light there was a search for water. It was a fruitless search, however, and the men became feverish, and, at last delirious, in which condition they abandoned their camp and wagons and separated into groups, walking over sands so hot as to burn their feet. When unable to proceed any further, they succumbed and their skeletons were found a few years later, bleached upon the sands.

Some of the stronger men in the party escaped the fate of their fellow travellers. It was on account of finding these bleached bones on the sands that the name of Death Valley was considered an appropriate one.

Since the early pioneers lost their lives in this gruesome desert, the place has been called "Death Valley" and the dread name has gone around the world.

It is in, and adjacent to, this barren, arid section of the United States, that one of the most useful chemicals in existance was discovered, BORAX, which has since blessed millions of American homes.

After the discovery of Borax the next question was to find a market for it. There was no prospect of ever getting a railroad into the great American desert, therefore other methods of transportation had to be secured. The 20 Mule Team solved the difficult problem, and thus the magic crystal that nature deposited in this barren desert, no one knows how many thousands of years ago, is found in almost every home.

Borax is recognized as one of the world's greatest cleansers. It is also an efficient antiseptic. It softens water better than any other known substance and not only cleanses thoroughly but purifies everything it comes in contact with. It is one of the mildest alkalies known. Millions of pounds of Borax are used annually by the British government to preserve mild cured hams. The Borax is dusted on the flesh side of the ham.

Send for a copy of our illustrated booklet on the uses of Borax in the home. It will be sent gratis to all who apply for it.

9 781015 845480